# The Marvelous Exploits of Paul Bunyan

## W.B. Laughead

# Contents

# THE MARVELOUS EXPLOITS
# OF PAUL BUNYAN

BY

W.B. Laughead

# Paul Bunyan
## Scholars Say He is the Only American Myth.

Paul Bunyan is the hero of lumbercamp whoppers that have been handed down for generations. These stories, never heard outside the haunts of the lumberjack until recent years, are now being collected by learned educators and literary authorities who declare that Paul Bunyan is "the only American myth."

The best authorities never recounted Paul Bunyan's exploits in narrative form. They made their statements more impressive by dropping them casually, in an off hand way, as if in reference. to actual events of common knowledge. To overawe the greenhorn in the bunkshanty, or the paper-collar stiffs and home guards in the saloons, a group of lumberjacks would remember meeting each other in the camps of Paul Bunyan. With painful accuracy they established the exact time and place, "on the Big Onion the winter of the blue snow" or "at Shot Gunderson's camp on the Tadpole the year of the sourdough drive." They elaborated on the old themes and new stories were born in lying contests where the heights of extemporaneous invention were reached.

In these conversations the lumberjack often took on the mannerisms of the French Canadian. This was apparently done without special intent and no reason for it can be given except for a similarity in the mock seriousness of their statements and the anti-climax of the bulls that were made, with the braggadocio of the habitant. Some investigators trace the origin of Paul Bunyan to Eastern Canada. Who can say?

Paul Bunyan came to Westwood, California, in 1913 at the suggestion of some of the most prominent loggers and lumbermen in the country. When the Red River Lumber Company announced their plans for opening up their forests of Sugar Pine and California White Pine, friendly advisors shook their heads and said,

"Better send for Paul Bunyan."

Apparently here was the job for a Superman, - quality-and-quantity-production on a big scale and great engineering difficulties to be overcome. Why not Paul Bunyan? This is a White Pine job and here in the High Sierras the winter snows lie deep, just like the country where Paul grew up. Here are trees that dwarf the largest "cork pine" of the Lake States and many new stunts were planned for logging, milling and manufacturing a product of supreme quality - just the job for Paul Bunyan.

The Red River people had been cutting White Pine in Minnesota for two generations; the crews that came west with them were old heads and every one knew Paul Bunyan of old. Paul had followed the White Pine from the Atlantic seaboard west to the jumping-off place in Minnesota, why not go the rest of the way?

Paul Bunyan's picture had never been published until he joined Red River and this likeness, first issued in 1914 is now the Red River trademark. It stands for the quality and service you have the right to expect from Paul Bunyan.

-

When and where did this mythical Hero get his start? Paul Bunyan is known by his mighty works, his antecedents and personal history are lost in doubt. You can prove that Paul logged off North Dakota and grubbed the stumps, not only by the fact that there are no traces of pine forests in that State, but by the testimony of oldtimers who saw it done. On the other hand, Paul's parentage and birth date are unknown. Like Topsy, he jes' growed.

Nobody cared to know his origin until the professors got after him. As long as he stayed around the camps his previous history was treated with the customary consideration and he was asked no questions, but when he broke into college it was all off. Then he had to have ancestors, a birthday and all sorts of vital statistics.

Now Paul is a regular myth and students of folklore make scientific research of "The Paul Bunyan Legend".

His first appearance in print was in the booklets published by The Red River Lumber Company in 1914 and 1916, these stories are reprinted in the present volume, with additions. Paul has followed the wanderings of pioneering workmen and performed new wonders in the oil fields, on big construction jobs and in the wheat fields but the stories in this book deal only with his work in the White Pine camps where he was born and raised. Care has been taken to preserve the atmosphere of the old style camps.

So now we will get on with Paul's doings and in the language of the four-horse skinner, "Let's dangle!"

Babe, the big blue ox constituted Paul Bunyan's assets and liabilities. History disagrees as to when, where and how Paul first acquired this bovine locomotive but his subsequent record is reliably established. Babe could pull anything that had two ends to it.

Babe was seven axehandles wide between the eyes according to some authorities; others equally dependable say forty-two axehandles and a plug of tobacco. Like other historical contradictions this comes from using different standards. Seven of Paul's axehandles were equal to a little more than forty-two of the ordinary kind.

When cost sheets were figured on Babe, Johnny Inkslinger found that upkeep and overhead were expensive but the charges for operation and depreciation were low and the efficiency was very high. How else could Paul have hauled logs to the landing a whole section (640 acres) at a time? He also used Babe to pull the kinks out of the crooked logging roads and it was on a job of this kind that Babe pulled a chain of three-inch links out into a straight bar.

They could never keep Babe more than one night at a camp for he would eat in one day all the feed one crew could tote to camp in a year. For a snack between meals he would eat fifty bales of hay, wire and all and six men with picaroons were kept busy picking the wire out of his teeth. Babe was a great pet and very docile as a general thing but he seemed to have a sense of humor and frequently got into mischief, He would sneak up behind a drive and drink all the water out of the river, leaving the logs high and dry. It was impossible to build an ox-sling big enough to hoist Babe off the ground for shoeing, but after they logged off Dakota there was room for Babe to lie down for this operation.

Once in a while Babe would run away and be gone all day roaming all over the Northwestern country. His tracks were so far apart that it was impossible to follow him and so deep that a man falling into one could only be hauled out with difficulty and a long rope. Once a settler and his wife and baby fell into one of these tracks and the son got out when he was fifty-seven years old and reported the accident. These tracks, today form the thousands of lakes in the "Land of the Sky-Blue Water."

Because he was so much younger than Babe and was brought to camp when a small calf, Benny was always called the Little Blue Ox although he was quite a chunk of an animal. Benny could not, or rather, would not haul as much as Babe nor was he as tractable but be could eat more.

Paul got Benny for nothing from a farmer near Bangor, Maine. There was not enough milk for the little fellow so he had to be weaned when three days old. The farmer only had forty acres of hay and by the time Benny was a week old he had to dispose of him for lack of food. The calf was undernourished and only weighed two tons when Paul got him. Paul drove from Bangor out to his headquarters camp near Devil's Lake, North Dakota that night and led Benny behind the sleigh. Western air agreed with the little calf and every time Paul looked back at him he was two feet taller.

When they arrived at camp Benny was given a good feed of buffalo milk and flapjacks and put into a barn by himself. Next morning the barn was gone. Later it was discovered on Benny's back as he scampered over the clearings. He had out-grown his barn in one night.

Benny was very notional and would never pull a load unless there was snow on the ground so after the spring thaws they had to white wash the logging roads to fool him.

Gluttony killed Benny. He had a mania for pancakes and one cook crew of two hundred men was kept busy making cakes for him. One night he pawed and bellowed and threshed his tail about till the wind of it blew down what pine Paul had left standing in Dakota. At breakfast time he broke loose, tore down the cook shanty and began bolting pancakes. In his greed he swallowed the red-hot stove. Indigestion set in and nothing could save him. What disposition was made of his body is a matter of dispute. One oldtimer claims that the outfit he works for bought a hind quarter of the carcass in 1857 and made corned beef of it. He thinks they have several carloads of it, left.

Another authority states that the body of Benny was dragged to a safe distance from the North Dakota camp and buried. When the earth was shoveled back it made a mound that formed the Black Hills in South Dakota.

The custodian and chaperon of Babe, the Big Blue Ox, was Brimstone Bill. He knew all the tricks of that frisky giant before they happened.

"I know oxen," the old bullwhacker used to say, "I've worked 'em and fed 'em and doctored 'em ever since the ox was invented. And Babe, I know that pernicious old reptyle same as if I'd abeen through him with a lantern."

Bill compiled "The Skinner's Dictionary," a hand book for teamsters, and most of the terms used in directing draft animals (except mules) originated with him. His early religious training accounts for the fact that the technical language of the teamster contains so many names of places and people spoken of in the Bible.

The buckskin harness used on Babe and Benny when the weather was rainy was made by Brimstone Bill. When this harness got wet it would stretch so much that the oxen could travel clear to the landing and the load would not move from the skidway in the woods. Brimstone would fasten the harness with an anchor Big Ole made for him and when the sun came out and the harness shrunk the load would be pulled to the landing while Bill and the oxen were busy at some other job.

The winter of the Blue Snow, the Pacific Ocean froze over and Bill kept the oxen busy hauling regular white snow over from China. M. H. Keenan can testify to the truth of this as he worked for Paul on the Big Onion that winter. It must have been about this time that Bill made the first ox yokes out of cranberry wood.

Feeding Paul Bunyan's crews was a complicated job. At no two camps were conditions the same. The winter he logged off North Dakota he had 300 cooks making pancakes for the Seven Axemen and the little Chore-boy. At headquarters on the Big Onion he had one cook and 462 cookees feeding a crew so big that Paul himself never knew within several hundred either way, how many men he had.

At Big Onion camp there was a lot of mechanical equipment and the trouble was a man who could handle the machinery cooked just like a machinist too. One cook got lost between the flour bin and the root cellar and nearly starved to death before he was found.

Cooks came and went. Some were good and others just able to get by. Paul never kept a poor one, very long. There was one jigger who seemed to have learned to do nothing but boil. He made soup out of everything and did most of his work with a dipper. When the big tote-sled broke through the ice on Bull Frog Lake with a load of split peas, he served warmed up, lake water till the crew struck. His idea of a lunch box was a jug or a rope to freeze soup onto like a candle. Some cooks used too much grease. It was said of one of these that he had to wear calked shoes to keep from sliding out of the cook-shanty and rub sand on his hands when he picked anything up.

There are two kinds of camp cooks, the Baking Powder Bums and the Sourdough Stiffs. Sourdough Sam belonged to the latter school. He made everything but coffee out of Sourdough. He had only one arm and one leg, the other members having been lost when his sourdough barrel blew up. Sam officiated at Tadpole River headquarters, the winter Shot Gunderson took charge.

After all others had failed at Big Onion camp, Paul hired his cousin Big Joe who came from three weeks below Quebec. This boy sure put a mean scald on the chuck. He was the only man who could make pancakes fast enough to feed the crew. He had Big Ole, the blacksmith, make him a griddle that was so big you couldn't see across it when the steam was thick. The batter, stirred in drums like concrete mixers was poured on with cranes and spouts. The griddle was greased by colored boys who skated over the surface with hams tied to their feet. They had to have colored boys to stand the heat.

At this camp the flunkeys wore roller skates and an idea of the size of the tables is gained from the fact that they distributed the pepper with four-horse teams.

Sending out lunch and timing the meals was rendered difficult by the size of the works which required three crews - one going to work, one on the job and one coming back. Joe had to start the bull-cook out with the lunch sled two weeks ahead of dinner time. To call the men who came in at noon was another problem. Big Ole made a dinner horn so big that no one could blow it but Big Joe or Paul himself. The first time Joe blew it be blew down ten acres of pine. The Red River people wouldn't stand for that so the next time he blew straight up but this caused severe cyclones and storms at sea so Paul had to junk the horn and ship it East where later it was made into a tin roof for a big Union Depot.

When Big Joe came to Westwood with Paul, he started something. About that time you may have read in the papers about a volcanic eruption at Mt. Lassen, heretofore extinct for many years. That was where Big Joe dug his bean-hole and when the steam worked out of the bean kettle and up through the ground, everyone thought the old hill had turned volcano. Every time Joe drops a biscuit they talk of earthquakes.

It was always thought that the quality of the food at Paul's Camps had a lot to do with the strength and endurance of the men. No doubt it did, but they were a husky lot to start with. As the feller said about fish for a brain food, "It won't do you no good unless there is a germ there to start with."

There must have been something to the food theory for the chipmunks that ate the prune pits got so big they killed all the wolves and years later the settlers shot them for tigers.

A visitor at one of Paul's camps was astonished to see a crew of men unloading four-horse logging sleds at the cook-shanty. They appeared to be rolling logs into a trap door from which poured clouds of steam.

"That's a heck of a place to land logs," he remarked.

"Them ain't logs," grinned a bull-cook, "them's sausages for the teamsters' breakfast."

At Paul's camp up where the little Gimlet empties into the Big Auger, new-comers used to kick because they were never served beans. The bosses and the men could never be interested in beans. E. E. Terrill tells us the reason:

Once when the cook quit they had to detail a substitute to the job temporarily. There was one man who was no good anywhere. He had failed at every job. Chris Crosshaul, the foreman, acting on the theory that every man is good somewhere, figured that this guy must be a cook, for it was the only job he had not tried. So he was put to work and the first thing he tackled was beans. He filled up a big kettle with beans and added some water. When the heat took hold the beans swelled up till they lifted off the roof and bulged out the walls. There was no way to get into the place to cook anything else, so the whole crew turned in to eat up the half cooked beans. By keeping at it steady they cleaned them up in a week and rescued the would-be-cook. After that no one seemed to care much for beans.

It used to be a big job to haul prune pits and coffee grounds away from Paul's camps. It required a big crew of men and either Babe or Benny to do the hauling. Finally Paul decided it was cheaper to build new camps and move every month.

The winter Paul logged off North Dakota with the Seven Axemen, the Little Chore Boy and the 300 cooks, he worked the cooks in three shifts - one for each meal. The Seven Axemen were hearty eaters; a portion of bacon was one side of a

1600-pound pig. Paul shipped a stern-wheel steamboat up Red River and they put it in the soup kettle to stir the soup.

Like other artists, cooks are temperamental and some of them are full of cussedness but the only ones who could sass Paul Bunyan and get away with it were the stars like Big Joe and Sourdough Sam. The lunch sled, - most popular institution in the lumber industry! Its arrival at, the noon rendezvous has been hailed with joy by hungry men on every logging job since Paul invented it. What if the warm food freezes on your tin plate, the keen cold air has sharpened your appetite to enjoy it. The crew that toted lunch for Paul Bunyan had so far to travel and so many to feed they hauled a complete kitchen on the lunch sled, cooks and all.

When Paul invented logging he had to invent all the tools and figure out all his own methods. There were no precedents. At the start his outfit consisted of Babe and his big axe.

No two logging jobs can be handled exactly the same way so Paul adapted his operations to local conditions. In the mountains he used Babe to pull the kinks out of the crooked logging roads; on the Big Onion he began the system of hauling a section of land at a time to the landings and in North Dakota he used the Seven Axemen.

At that time marking logs was not thought of, Paul had no need for identification when there were no logs but his own. About the time he started the Atlantic Ocean drive others had come into the industry and although their combined cut was insignificant compared to Paul's, there was danger of confusion, and Paul had most to lose.

At first Paul marked his logs by pinching a piece out of each log. When his cut grew so large that the marking had to be detailed to the crews, the "scalp" on each log was put on with an axe, for even in those days not every man could nip out the chunk with his fingers.

The Grindstone was invented by Paul the winter he logged off North Dakota. Before that Paul's axemen had to sharpen their axes by rolling rocks down hill and running along side of them. When they got to "Big Dick," as the lumberjacks called Dakota, hills and rocks were so hard to find that Paul rigged up the revolving rock.

This was much appreciated by the Seven Axemen as it enabled them to grind an axe in a week, but the grindstone was not much of a hit with the Little Chore Boy whose job it was to turn it. The first stone was so big that working at full speed, every time it turned around once it was payday.

The Little Chore Boy led a strenuous life. He was only a kid and like all young-sters putting in their first winter in the woods, he was put over the jumps by the oldtimers. His regular work was heavy enough, splitting all the wood for the camp, carrying water and packing lunch to the men, but his hazers sent him on all kinds of wild goose errands to all parts of the works, looking for a "left-handed peavy" or a "bundle of cross-hauls."

He had to take a lot of good natured roughneck wit about his size for he only weighed 800 pounds and a couple of surcingles made a belt for him. What he lacked in size he made up in grit and the men secretly respected his gameness. They said he might make a pretty good man if he ever got any growth, and considered it a necessary education to give him a lot of extra chores.

Often in the evening, after his day's work and long hours put in turning the grindstone and keeping up fires in the camp stoves - that required four cords of wood apiece to kindle a fire, he could be found with one of Big Ole's small 600-pound anvils in his lap pegging up shoes with railroad spikes.

It was a long time before they solved the problem of turning logging sleds around in the road. When a sled returned from the landing and put on a load they had to wait until Paul came along to pick up the four horses and the load and head them the other way. Judson M. Goss says he worked for Paul the winter he in-vented the round turn.

All of Paul's inventions were successful except when he decided to run three ten-hour shifts a day and installed the Aurora Borealis. After a number of trials the plan was abandoned because the lights were not dependable.

"The Seven Axemen of the Red River" they were called because they had a camp on Red River with the three-hundred cooks and the Little Chore Boy. The whole State was cut over from the one camp and the husky seven chopped from dark to dark and walked to and from work.

Their axes were so big it took a week to grind one of them. Each man had three axes and two helpers to carry the spare axes to the river when they got red hot from chopping. Even in those days they had to watch out for forest fires. The axes were hung on long rope handles. Each axeman would march through the timber whirling his axe around him till the hum of it sounded like one of Paul's for-and-aft mosquitoes, and at every step a quarter-section of timber was cut.

The height, weight and chest measurement of the Seven Axemen are not known. Authorities differ. History agrees that they kept a cord of four-foot wood on the table for toothpicks. After supper they would sit on the deacon seat in the bunk shanty and sing "Shanty Boy" and "Bung Yer Eye" till the folks in the settlements down on the Atlantic would think another nor'wester was blowing up.

Some say the Seven Axemen were Bay Chaleur men; others declare they were all cousins and came from down Machias way. Where they came from or where they went to blow their stake after leaving Paul's camp no one knows but they are remembered as husky lads and good fellows around camp.

After the Seven Axemen had gone down the tote road, never to return, Paul Bunyan was at a loss to find a method of cutting down trees that would give him anything like the output he had been getting. Many trials and experiments followed and then Paul invented the two-man Saw.

The first saw was made from a strip trimmed off in making Big Joe's dinner horn and was long enough to reach across a quarter section, for Paul could never think in smaller units. This saw worked all right in a level country, in spite of the fact that all the trees fell back on the saw, but in rough country only the trees on the hill tops were cut. Trees in the valleys were cut off in the tops and in the pot holes the saw passed over the trees altogether.

It took a good man to pull this saw in heavy timber when Paul was working on the other end. Paul used to say to his fellow sawyer, "I don't care if you ride the saw, but please don't drag your feet." A couple of cousins of Big Ole's were given the job and did so well that ever afterward in the Lake States the saw crews have generally been Scandinavians.

It was after this that Paul had Big Ole make the "Down-Cutter." This was a rig like a mowing machine. They drove around eight townships and cut a swath 500 feet wide.

Paul Bunyan's Trained Ants are proving so successful that they may replace donkeys and tractors on the rugged slopes of the Sierras. Inspired by his success with Bees and Mosquitoes, Paul has developed a breed of Ants that stand six feet tall and weigh 200 pounds.

To overcome their habit of hibernating all Winter, Paul supplied the Ants with Mackinaws made with three pairs of sleeves or legs. They eat nothing but Copenhagen Snuff. The Ants (or Uncles as they prefer to be called) can run to the Westwood shops with a damaged locomotive quicker than the Wrecking Crew can come out. They do not patronize bootleggers or require time off to fix their automobiles.

Lucy, Paul Bunyan's cow was not, so far as we can learn, related to either Babe or Benny. Statements that she was in any way their mother are without basis in fact. The two oxen had been in Paul's possession for a long time before Lucy arrived on the scene.

No reliable data can be found as to the pedigree of this remarkable dairy animal. There are no official records of her butterfat fat production nor is it known where or how Paul got her.

Paul always said that Lucy was part Jersey and part wolf. Maybe so. Her actions and methods of living seemed to justify the allegation of wolf ancestry, for she had an insatiable appetite and a roving disposition. Lucy ate everything in sight and could never be fed at the same camp with Babe or Benny. In fact, they quit trying to feed her at all but let her forage her own living. The Winter of the Deep Snow, when even the tallest White Pines were buried, Brimstone Bill outfitted Lucy with a set of Babe's old snowshoes and a pair of green goggles and turned her out to graze on the snowdrifts. At first she had some trouble with the new foot gear but once she learned to run them and shift gears without wrecking herself, she answered the call of the limitless snow fields and ran away all over North America until Paul decorated her with a bell borrowed from a buried church.

In spite of short rations she gave enough milk to keep six men busy skimming the cream. If she bad been kept in a barn and fed regularly she might have made a milking record. When she fed on the evergreen trees and her milk got so strong of White Pine and Balsam that the men used it for cough medicine and liniment, they quit serving the milk on the table and made butter out of it. By using this butter, to grease the logging roads when the snow and ice thawed off, Paul was able to run big logging sleds all summer.

The family life of Paul Bunyan, from all accounts, has been very happy. A charming glimpse of Mrs. Bunyan is given by Mr. E. S. Shepard of Rhinelander,

Wis., who tells of working in Paul's camp on Round River in '62, the Winter of the Black Snow. Paul put him wheeling prune pits away from the cook camp. After he had worked at this job for three months Paul had him haul them back again as Mrs. Bunyan, who was cooking at the camp, wanted to use them to make the hot fires necessary to cook her famous soft nosed pancakes.

Mrs. Bunyan, at this time used to call the men to dinner by blowing into a woodpecker hole in an old hollow stub that stood near the door. In this stub there was a nest of owls that had one short wing and flew in circles. When Mr. Shepard made a sketch of Paul, Mrs. Bunyan, with wifely solicitude for his appearance, parted Paul's hair with a handaxe and combed it with an old cross-cut saw.

From other sources we have fragmentary glimpses of Jean, Paul's youngest son. When Jean was three weeks old he jumped from his cradle one night and seizing an axe, chopped the four posts out from under his father's bed. The incident greatly tickled Paul, who used to brag about it to any one who would listen to him. "The boy is going to be a great logger some day," he would declare with fatherly pride.

The last we heard of Jean he was working for a lumber outfit in the South, lifting logging trains past one another on a single track railroad.

What is camp without a dog? Paul Bunyan loved dogs as well as the next man but never would have one around that could not earn its keep. Paul's dogs had to work, hunt or catch rats. It took a good dog to kill the rats and mice in Paul's camp for the rodents picked up scraps of the buffalo milk pancakes and grew to be as big as two year old bears.

Elmer, the moose terrier, practiced up on the rats when he was a small pup and was soon able to catch a moose on the run and finish it with one shake. Elmer loafed around the cook camp and if the meat supply happened to run low the cook would put the dog out the door and say, "Bring in a moose." Elmer would run into the timber, catch a moose and bring it in and repeat the performance until, after a few minutes work, the cook figured he had enough for a mess and would call the dog in.

Sport, the reversible dog was really the best hunter. He was part wolf and part elephant hound and was raised on bear milk. One night when Sport was quite young, he was playing around in the horse barn and Paul, mistaking him for a mouse, threw a band axe at him. The axe cut the dog in two but Paul, instantly realizing what had happened, quickly stuck the two halves together, gave the pup first aid and bandaged him up. With careful nursing the dog soon recovered and then it was seen that Paul in his haste had twisted the two halves so that the hind legs pointed straight up. This proved to be an advantage for the dog learned to run on one pair of legs for a while and then flop over without loss of speed and run on the other pair. Because of this he never tired and anything he started after got caught. Sport never got his full growth. While still a pup he broke through four feet of ice on Lake Superior and was drowned.

As a hunter, Paul would make old Nimrod himself look like a city dude lost from his guide. He was also a good fisherman. Old-timers tell of seeing Paul as a small boy, fishing off the Atlantic Coast. He would sail out early in the morning in his three-mast schooner and wade back before breakfast with his boat full of fish on his shoulder.

About this time he got his shot gun that required four dishpans full of powder and a keg of spikes to load each barrel. With this gun he could shoot geese so high in the air they would spoil before reaching the ground.

Tracking was Paul's favorite sport and no trail was too old or too dim for him to follow. He once came across the skeleton of a moose that had died of old age and, just for curiosity, picked up the tracks of the animal and spent the whole afternoon following its trail back to the place where it was born.

The shaggy dog that spent most of his time pretending to sleep in front of Johnny Inkslinger's counter in the camp office was Fido, the watch dog. Fido was the bug-bear (not bearer, just bear) of the greenhorns. They were told that Paul starved Fido all winter and then, just before payday, fed him all the swampers, barn boys, and student bullcooks. The very marrow was frozen in their heads at the thought of being turned into dog food. Their fears were groundless for Paul would never let a dog go hungry or mistreat a human being. Fido was fed all the watch peddlers, tailors' agents, and camp inspectors and thus served a very useful purpose.

It is no picnic to tackle the wilderness and turn the very forest itself into a commercial commodity delivered at the market. A logger needs plenty of brains and back bone.

Paul Bunyan had his setbacks the same as every logger only his were worse. Being a pioneer he had to invent all his stuff as he went along. Many a time his plans were upset by the mistakes of some swivel-headed strawboss or incompetent fore-man. The winter of the blue snow, Shot Gunderson had charge in the Big Tadpole River country. He landed all of his logs in a lake and in the spring when ready to drive he boomed the logs three times around the lake before be discovered there was no outlet to it. High hills surrounded the lake and the drivable stream was ten miles away. Apparently the logs were a total loss.

Then Paul came on the job himself and got busy. Calling in Sourdough Sam, the cook who made everything but coffee out of sourdough, he ordered him to mix enough sourdough to fill the big watertank. Hitching Babe to the tank he hauled it over and dumped it into the lake. When it "riz," as Sam said, a mighty lava-like stream poured forth and carried the logs over the hills to the river. There is a land-locked lake in Northern Minnesota that is called "Sourdough Lake" to this day.

Chris Crosshaul was a careless cuss. He took a big drive down the Mississippi for Paul and when the logs were delivered in the New Orleans boom it was found that he had driven the wrong logs. The owners looked at the barkmarks and refused to accept them. It was up to Paul to drive them back upstream.

No one but Paul Bunyan would ever tackle a job like that. To drive logs up-stream is impossible, but if you think a little thing like an impossibility could stop him, you don't know Paul Bunyan. He simply fed Babe a good big salt ration and drove him to the upper Mississippi to drink. Babe drank the river dry and sucked all the water upstream. The logs came up river faster than they went down.

Big Ole was the Blacksmith at Paul's headquarters camp on the Big Onion. Ole had a cranky disposition but he was a skilled workman. No job in iron or steel was too big or too difficult for him. One of the cooks used to make doughnuts and have Ole punch the holes. He made the griddle on which Big Joe cast his pancakes and the dinner horn that blew down ten acres of pine. Ole was the only man who could shoe Babe or Benny. Every time he made a set of shoes for Babe they had to open up another Minnesota iron mine. Ole once carried a pair of these shoes a mile and sunk knee deep into solid rock at every step. Babe cast a shoe while making a hard pull one day, and it was hurled for a mile and tore down forty acres of pine and in-jured eight Swedes that were swamping out skidways. Ole was also a mechanic and built the Downcutter, a rig like a mowing machine that cut down a swath of trees

500 feet wide.

-

In the early days, whenever Paul Bunyan was broke between logging seasons, he traveled around like other lumberjacks doing any kind of pioneering work he could find. He showed up in Washington about the time The Puget Construction Co. was building Puget Sound and Billy Puget was making records moving dirt with droves of dirt throwing badgers. Paul and Billy got into an argument over who had shoveled the most. Paul got mad and said he'd show Billy Puget and started to throw the dirt back again. Before Billy stopped him he had piled up the San Juan Islands.

When a man gets the reputation in the woods of being a "good man" it refers only to physical prowess. Frequently he is challenged to fight by "good men" from other communities.

There was Pete Mufraw. "You know Joe Mufraw?" "Oui, two Joe Mufraw, one named Pete." That's the fellow. After Pete had licked everybody between Quebec and Bay Chaleur he started to look for Paul Bunyan. He bragged all over the country that he had worn out six pair of shoe-pacs looking for Paul. Finally he met up with him.

Paul was plowing with two yoke of steers and Pete Mufraw stopped at the brush-fence to watch the plow cut its way right through rocks and stumps. When they reached the end of the furrow Paul picked up the plow and the oxen with one arm and turned them around. Pete took one look and then wandered off down the trail muttering, "Hox an' hall! She's lift hox an' hall."

Paul Bunyan started traveling before the steam cars were invented. He developed his own means of transportation and the railroads have never been able to catch up. Time is so valuable to Paul he has no time to fool around at sixty miles an hour.

In the early days he rode on the back of Babe, the Big Blue Ox. This had its difficulties because he had to use a telescope to keep Babe's hind legs in view and the hooves of the ox created such havoc that after the settlements came into different parts of the country there were heavy damage claims to settle every trip.

Snowshoes were useful in winter but one trip on the webs cured Paul of depending upon them for transcontinental bikes. He started from Minnesota for

Westwood one Spring morning. There was still snow in the woods so Paul wore his snowshoes. He soon ran out of the snow belt but kept right on without reducing speed. Crossing the desert the heat became oppressive, his mackinaws grew heavy and the snowshoes dragged his feet but it was too late to turn back.

When he arrived in California he discovered that the sun and hot sand had warped one of his shoes and pulled one foot out of line at every step, so instead of traveling on a bee line and hitting Westwood exactly, he came out at San Francisco. This made it necessary for him to travel an extra three hundred miles north. It was late that night when he pulled into Westwood and he had used up a whole day coming from Minnesota.

Paul's fast foot work made him a "good man on the round stuff" and in spite of his weight he had no trouble running around on the floating logs, even the small ones. It was said that Paul could spin a log till the bark came off and then run ashore on the bubbles. He once threw a peavy handle into the Mississippi at St. Louis and standing on it, poled up to Brainerd, Minnesota. Paul was a "white water bucko" and rode water so rough it would tear an ordinary man in two to drink out of the river.

Johnny Inkslinger was Paul's headquarters clerk. He invented bookkeeping about the time Paul invented logging. He was something of a genius and perfected his own office appliances to increase efficiency. His fountain pen was made by running a hose from a barrel of ink and with it he could "daub out a walk" quicker than the recipient of the pay-off could tie the knot in his tussick rope.

One winter Johnny left off crossing the "t's" and dotting the "i's" and saved nine barrels of ink. The lumberjacks accused him of using a split pencil to charge up the tobacco and socks they bought at the wanagan but this was just bunkshanty talk (is this the origin of the classic term "the bunk"?) for Johnny never cheated anyone.

Have you ever encountered the Mosquito of the North Country? You thought they were pretty well developed animals with keen appetites, didn't you? Then you can appreciate what Paul Bunyan was up against when he was surrounded by the vast swarms of the giant ancestors of the present race of mosquitoes, getting their first taste of human victims. The present mosquito is but a degenerate remnant of the species. Now they rarely weigh more than a pound or measure more than fourteen or fifteen inches from tip to tip.

Paul had to keep his men and oxen in the camps with doors and windows barred. Men armed with pikepoles and axes fought off the insects that tore the shakes off the roof in their efforts to gain entrance. The big buck mosquitoes fought among themselves and trampled down the weaker members of the swarm and to this alone Paul Bunyan and his crew owe their lives.

Paul determined to conquer the mosquitoes before another season arrived. He thought of the big Bumble Bees back home and sent for several yoke of them. These, he hoped would destroy the mosquitoes. Sourdough Sam brought out two pair of bees, overland on foot. There was no other way to travel for the flight of the beasts could not be controlled. Their wings were strapped with surcingles, they checked their stingers with Sam and walking shoes were provided for them. Sam brought them through without losing a bee.

The cure was worse than the original trouble. The Mosquitoes and the Bees made a hit with each other. They soon intermarried and their off-spring, as often happens, were worse than their parents. They had stingers fore-and-aft and could get you coming or going.

Their bee blood caused their downfall in the long run. Their craving for sweets could only be satisfied by sugar and molasses in large quantities, for what is a flower to an insect with a ten-gallon stomach? One day the whole tribe flew across Lake Superior to attack a fleet of ships bringing sugar to Paul's camps. They destroyed the ships but ate so much sugar they could not fly and all were drowned.

One pair of the original bees were kept at headquarters camp and provided honey for the pancakes for many years.

-

If Paul Bunyan did not invent Geography he created a lot of it. The Great Lakes were first constructed to provide a water hole for Babe the Big Blue Ox. Just what year his work was done is not known but they were in use prior to the Year of the Two Winters.

The Winter Paul Bunyan logged off North Dakota he hauled water for his ice roads from the Great Lakes. One day when Brimstone Bill had Babe hitched to one of the old water tanks and was making his early morning trip, the tank sprung a leak when they were half way across Minnesota. Bill saved himself from drowning by climbing Babe's tail but all efforts to patch up the tank were in vain so the old tank was abandoned and replaced by one of the new ones. This was the beginning of the Mississippi River and the truth of this is established by the fact that the old Mississippi is still flowing.

The cooks in Paul's camps used a lot of water and to make things handy, they used to dig wells near the cook shanty. At headquarters on the Big Auger, on top of the hill near the mouth of the Little Gimlet, Paul dug a well so deep that it took all day for the bucket to fall to the water, and a week to haul it up. They had to run so many buckets that the well was forty feet in diameter. It was shored up with tamarac poles and when the camp was abandoned Paul pulled up this cribbing. Travelers who have visited the spot say that the sand has blown away until 178 feet of the well is sticking up into the air, forming a striking landmark.

The Winter of the Deep Snow everything was buried. Paul had to dig down to find the tops of the tallest White Pines. He had the snow dug away around them and lowered his sawyers down to the base of the trees. When the tree was cut off he hauled it to the surface with a long parbuckle chain to which Babe, mounted on snowshoes, was hitched. It was impossible to get enough stove pipe to reach to the top of the snow, so Paul had Big Ole make stovepipe by boring out logs with a long six-inch auger.

The year of the Two Winters they had winter all summer and then in the fall it turned colder. One day Big Joe set the boiling coffeepot on the stove and it froze so quick that the ice was hot. That was right after Paul had built the Great Lakes and that winter they froze clear to the bottom. They never would have thawed out if Paul had not chopped out the ice and hauled it out on shore for the sun to melt. He finally got all the ice thawed but he had to put in all new fish.

The next spring was the year the rain came up from China. It rained so hard and so long that the grass was all washed out by the roots and Paul had a great time feeding his cattle. Babe had to learn to eat pancakes like Benny. That was the time Paul used the straw hats for an emergency ration.

When Paul's drive came down, folks in the settlements were astonished to see all the river-pigs wearing huge straw bats. The reason for this was soon apparent. When the fodder ran out every man was politely requested to toss his hat into the ring. Hundreds of straw hats were used to make a lunch for Babe.

When Paul Bunyan took up efficiency engineering he went at the the job with all his customary thoroughness. He did not fool around clocking the crew with a stop watch, counting motions and deducting the ones used for borrowing chews, going for drinks, dodging the boss and preparing for quitting time. He decided to cut out labor altogether.

"What's the use," said Paul, "of all this sawing, swamping, skidding, decking, grading and icing roads, loading, hauling and landing? The object of the game is to get the trees to the landing, ain't it? Well, why not do it and get it off your mind?"

So he hitched Babe to a section of land and snaked in the whole 640 acres at one drag. At the landing the trees were cut off just like shearing a sheep and the denuded section hauled back to its original place. This simplified matters and made the work a lot easier. Six trips a day, six days a week just cleaned up a township for section 37 was never hauled back to the woods on Saturday night but was left on the landing to wash away in the early spring when the drive went out,

Documentary evidence of the truth of this is offered by the United States government surveys. Look at any map that shows the land subdivisions and you will never find a township with more than thirty-six sections.

The foregoing statement, previously published, has caused some controversy. Mr. T. S. Sowell of Miami, Florida wrote to us citing the townships in his State that have sections numbered 37 to 40. He said that the government survey had been complicated by the old Spanish land grants. We put the matter up to Paul Bunyan and from his camp near Westwood came this reply:

Red River Advertising Department.

Dear Sir: Yes sir, I remember those sections and a lot of bother they made me too. One winter when I was starting the White Pine business and snaking sections down to the Atlantic Ocean, a man from Florida came along and ordered a bunch of sections delivered down to his place. He wanted to see if he could grow the same kind of White Pine down there. I yarded out a nice bunch of sections and next summer when my drive was in and I wasn't busy I took a crew of Canada Boys and Mainites and poled them down the coast. When I come to collect they said this man was gone looking for a Fountain of Youth or some fool thing.

I don't know what luck he had with his White Pine ranch. I never seen them again. I had a lot of other things to tend to and clean forgot it till you sent me Mr.

Sowell's letter. Maybe that man was a Spaniard I don't know.

Yours respectively, P. Bunyan.

-

From 1917 to 1920 Paul Bunyan was busy toting the supplies and building camps for a bunch of husky young fellow-Americans who bad a contract on the other side of the Atlantic, showing a certain prominent European (who is now logging in Holland) how they log in the United States.

After his service overseas with the A. E. F., Paul couldn't get back to the States quick enough. Airplanes were too slow so Paul embarked in his Bark Canoe, the one he used on the Big Onion the year he drove logs upstream. When be threw the old paddle into high he sure rambled and the sea was covered with dead fish that broke their backs trying to watch him coming and going.

As he shoved off from France, Paul sent a wireless to New York but passed the Statue of Liberty three lengths ahead of the message. From New York to Westwood he traveled on skis. When the home folks asked him if the Allegheney Mountains and the Rockies had bothered him, Paul replied, "I didn't notice any mountains but the trail was a little bumpy in a couple of spots."

In the forests of the Red River Lumber Company Paul Bunyan can cut his lumber for many future years in the region where Nature found conditions exactly suited to the growth of pine of the finest texture and largest size.

Early in the closing decade of the nineteenth century the Red River people took a long look into the future. Foreseeing the exhaustion of their Minnesota white pine, which came a quarter of a century later, they set out to find the pine that would take its place. Their search covered several years and reached all the important stands in the western States. This was well in advance of the westward movement of the industry and Red River had the pioneer's opportunity for choice and rejection.

Sugar Pine, "cork pine's big brother," is botanically and physically true white pine, with all the family virtues. It is the largest of all pines.

California Pine is the trade name for pinus ponderosa or western yellow pine from certain regions where conditions of growth have so modified the nature of the wood that it is more like white pine than it is like its botanical brothers that grow elsewhere. Some say this change is due to volcanic soil. Whatever the cause, California Pine from Red River's forest is exceptionally light, brightly colored, soft and even textured and second only to Sugar Pine in size.

Red River "Paul Bunyan's" California Pine and Sugar Pine meet the strict requirements of trades that have made white pine their standard. Where freedom from distortion is essential, as for example piano actions, organ pipes, foundry patterns and the best sash and doors, Red River pines are used. They finish economically with paints, stains and enamels and are highly valued as cores for fine hardwood veneers. They work easily, smoothly and cleanly with edged tools and do not nail-split.

The durability of these California pines is shown by their sound condition in California buildings that have stood for generations, many of them in regions where climatic conditions are more conducive to decay than in the middle western and eastern states.

Paul Bunyan tackled a real problem when he came to Westwood. The site of the mill and town was unbroken forest in 1913, sixty mountainous miles from the nearest railroad. Trails were graded into passable roads and materials and machinery were freighted in. When the railroad arrived in 1914 the first mill was in operation and the town well under construction. Town and plant had been detailed on the drafting boards in Minneapolis. Sanitary sewers, water system, electric lights and telephones were extended as the forest was cleared and Westwood, with a population of 5,000, enjoys all the facilities of a modern American community.

The electrically operated sawmill has an annual capacity of 250 million board feet. Dry kilns, one of the largest plywood factories in the country, sash and door factory and re-manufacturing departments round out production of a complete line of lumber products.

Red River operates its own logging railroad, 20 miles of which are electrified, hydro-electric plants and the foundry and machine shops, where many units of the logging and plant machinery are designed and built.

Back in the early days, when his camps were so far from any where that the wolves following the tote-teams got lost in the woods, Paul Bunyan made no attempt to keep in touch with the trade. What's the use when every letter that comes in is about things that happened the year before?

Since he came to Westwood Paul has renewed old friendships, formed new ones and kept close contact with the world. Everyone expects great things of Paul Bunyan and with the Red River outfit back of him he has the chance of his life to make good. Continuous production keeps a full assortment of stock on hand. Customers in all parts of America find Westwood a dependable source of supply.

Here is an instance. This old friend of Paul's a prominent furniture manufacturer in the Lake States, was disappointed because an item he wanted for immediate shipment was not in stock in the grade and thickness required. He wrote the letter shown below and was given an explanation of the facts in the case in the accompanying reply.

Paul Bunyan Makes Plywood

Paul Bunyan says that making plywood reminds him of the way Mrs. Bunyan made pies during the hard times of pioneer days. She would take pancakes, spread molasses between and sew around the edges with yarn.

Plywood panels differ from other wall coverings in that the natural texture of the wood is not altered. While the lathe-cut sheets are thin, they are solid wood with the cell structure just the same as it grew in the tree. In making plywood the inside sheets are placed crossgrained with the face sheets. These sheets are then united with a glue bond that is stronger than the wood itself. This cross-grained construction prevents splitting and produces a panel much stronger than solid wood of the same thickness.

Paul Bunyan's California Pines give Red River plywood's a distinctive character. They carry the qualities that have given "old-fashioned white pine" its long-established preference by craftsmen and builders. The soft, even texture takes up paints, stains and enamels economically and gives a fine finish, unmarred by checking and "grainraising" when properly handled.

Red River construction embodies special features in the process of re-drying and in cutting for straight grain. The latest and best developments in the manufacture of glues and in their scientific application are utilized. Painstaking workmanship and careful inspection and grading make Red River plywood's outstanding in quality.

Plywood panels have revolutionized the use of wood in building and in industry. From the growing list of industrial uses we might note the following as typical: trunks, concrete forms, furniture backs, drawer bottoms and cores for fine hardwood veneers; cabinets, car bodies, boxes, table and counter tops, door panels, signs, toys and ship bulkheads.

Builders use plywood panels for interior walls and ceilings and for insulation, sub-floors, sheathing, shelving, cupboards and built-in units. The richness of wood-paneled rooms can now be enjoyed at a cost that compares favorably with other wall coverings. The paneled interiors do not go out of style or require redecoration. They are not damaged by water or shock and ordinary breakage. They do not crack or peel.

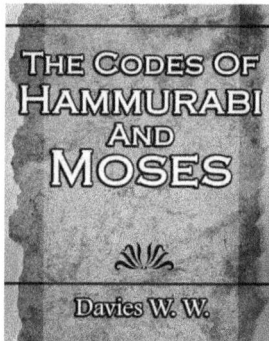

### The Codes Of Hammurabi And Moses
### W. W. Davies

QTY

The discovery of the Hammurabi Code is one of the greatest achievements of archaeology, and is of paramount interest, not only to the student of the Bible, but also to all those interested in ancient history...

**Religion**    **ISBN:** *1-59462-338-4*    **Pages:132**

*MSRP $12.95*

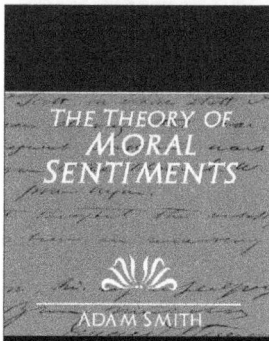

### The Theory of Moral Sentiments
### Adam Smith

QTY

This work from 1749. contains original theories of conscience amd moral judgment and it is the foundation for systemof morals.

**Philosophy**   **ISBN:** *1-59462-777-0*    **Pages:536**

*MSRP $19.95*

### Jessica's First Prayer
### Hesba Stretton

QTY

In a screened and secluded corner of one of the many railway-bridges which span the streets of London there could be seen a few years ago, from five o'clock every morning until half past eight, a tidily set-out coffee-stall, consisting of a trestle and board, upon which stood two large tin cans, with a small fire of charcoal burning under each so as to keep the coffee boiling during the early hours of the morning when the work-people were thronging into the city on their way to their daily toil...

**Pages:84**

**Childrens**   **ISBN:** *1-59462-373-2*    *MSRP $9.95*

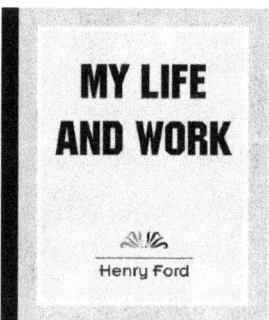

### My Life and Work
### Henry Ford

QTY

Henry Ford revolutionized the world with his implementation of mass production for the Model T automobile. Gain valuable business insight into his life and work with his own auto-biography... "We have only started on our development of our country we have not as yet, with all our talk of wonderful progress, done more than scratch the surface. The progress has been wonderful enough but..."

**Pages:300**

**Biographies/**   **ISBN:** *1-59462-198-5*    *MSRP $21.95*

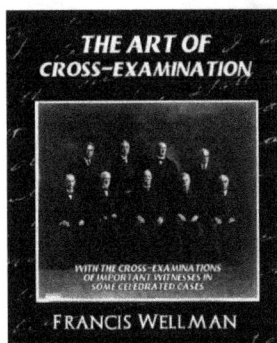

## The Art of Cross-Examination
## Francis Wellman

QTY

I presume it is the experience of every author, after his first book is published upon an important subject, to be almost overwhelmed with a wealth of ideas and illustrations which could readily have been included in his book, and which to his own mind, at least, seem to make a second edition inevitable. Such certainly was the case with me; and when the first edition had reached its sixth impression in five months, I rejoiced to learn that it seemed to my publishers that the book had met with a sufficiently favorable reception to justify a second and considerably enlarged edition. ..

**Pages:412**

**Reference**   **ISBN:** *1-59462-647-2*   *MSRP $19.95*

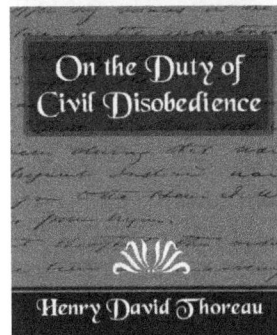

## On the Duty of Civil Disobedience
## Henry David Thoreau

QTY

Thoreau wrote his famous essay, On the Duty of Civil Disobedience, as a protest against an unjust but popular war and the immoral but popular institution of slave-owning. He did more than write—he declined to pay his taxes, and was hauled off to gaol in consequence. Who can say how much this refusal of his hastened the end of the war and of slavery ?

**Law**      **ISBN:** *1-59462-747-9*      **Pages:48**

*MSRP $7.45*

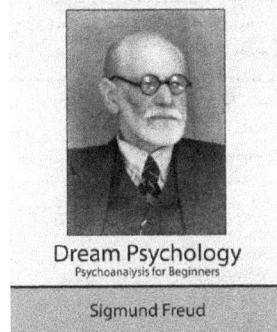

## Dream Psychology Psychoanalysis for Beginners
## Sigmund Freud

QTY

Sigmund Freud, born Sigismund Schlomo Freud (May 6, 1856 - September 23, 1939), was a Jewish-Austrian neurologist and psychiatrist who co-founded the psychoanalytic school of psychology. Freud is best known for his theories of the unconscious mind, especially involving the mechanism of repression; his redefinition of sexual desire as mobile and directed towards a wide variety of objects; and his therapeutic techniques, especially his understanding of transference in the therapeutic relationship and the presumed value of dreams as sources of insight into unconscious desires.

**Pages:196**

**Psychology**   **ISBN:** *1-59462-905-6*   *MSRP $15.45*

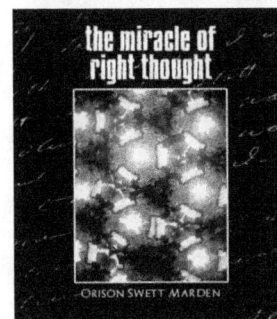

## The Miracle of Right Thought
## Orison Swett Marden

QTY

Believe with all of your heart that you will do what you were made to do. When the mind has once formed the habit of holding cheerful, happy, prosperous pictures, it will not be easy to form the opposite habit. It does not matter how improbable or how far away this realization may see, or how dark the prospects may be, if we visualize them as best we can, as vividly as possible, hold tenaciously to them and vigorously struggle to attain them, they will gradually become actualized, realized in the life. But a desire, a longing without endeavor, a yearning abandoned or held indifferently will vanish without realization.

**Pages:360**

**Self Help**      **ISBN:** *1-59462-644-8*      *MSRP $25.45*

**QTY**

**The Rosicrucian Cosmo-Conception Mystic Christianity** *by Max Heindel*    ISBN: *1-59462-188-8*   **$38.95**
*The Rosicrucian Cosmo-conception is not dogmatic, neither does it appeal to any other authority than the reason of the student. It is: not controversial, but is: sent forth in the, hope that it may help to clear...*    New Age/Religion Pages 646

**Abandonment To Divine Providence** *by Jean-Pierre de Caussade*    ISBN: *1-59462-228-0*   **$25.95**
*"The Rev. Jean Pierre de Caussade was one of the most remarkable spiritual writers of the Society of Jesus in France in the 18th Century. His death took place at Toulouse in 1751. His works have gone through many editions and have been republished...*    Inspirational/Religion Pages 400

**Mental Chemistry** *by Charles Haanel*    ISBN: *1-59462-192-6*   **$23.95**
*Mental Chemistry allows the change of material conditions by combining and appropriately utilizing the power of the mind.  Much like applied chemistry creates something new and unique out of careful combinations of chemicals the mastery of mental chemistry...*    New Age Pages 354

**The Letters of Robert Browning and Elizabeth Barret Barrett 1845-1846 vol II**    ISBN: *1-59462-193-4*   **$35.95**
*by Robert Browning and Elizabeth Barrett*    Biographies Pages 596

**Gleanings In Genesis (volume I)** *by Arthur W. Pink*    ISBN: *1-59462-130-6*   **$27.45**
*Appropriately has Genesis been termed "the seed plot of the Bible" for in it we have, in germ form, almost all of the great doctrines which are afterwards fully developed in the books of Scripture which follow...*    Religion/Inspirational Pages 420

**The Master Key** *by L. W. de Laurence*    ISBN: *1-59462-001-6*   **$30.95**
*In no branch of human knowledge has there been a more lively increase of the spirit of research during the past few years than in the study of Psychology, Concentration and Mental Discipline. The requests for authentic lessons in Thought Control, Mental Discipline and...*    New Age/Business Pages 422

**The Lesser Key Of Solomon Goetia** *by L. W. de Laurence*    ISBN: *1-59462-092-X*   **$9.95**
*This translation of the first book of the "Lemegton" which is now for the first time made accessible to students of Talismanic Magic was done, after careful collation and edition, from numerous Ancient Manuscripts in Hebrew, Latin, and French...*    New Age/Occult Pages 92

**Rubaiyat Of Omar Khayyam** *by Edward Fitzgerald*    ISBN: *1-59462-332-5*   **$13.95**
*Edward Fitzgerald, whom the world has already learned, in spite of his own efforts to remain within the shadow of anonymity, to look upon as one of the rarest poets of the century, was born at Bredfield, in Suffolk, on the 31st of March, 1809. He was the third son of John Purcell...*    Music Pages 172

**Ancient Law** *by Henry Maine*    ISBN: *1-59462-128-4*   **$29.95**
*The chief object of the following pages is to indicate some of the earliest ideas of mankind, as they are reflected in Ancient Law, and to point out the relation of those ideas to modern thought.*    Religion/History Pages 452

**Far-Away Stories** *by William J. Locke*    ISBN: *1-59462-129-2*   **$19.45**
*"Good wine needs no bush, but a collection of mixed vintages does. And this book is just such a collection. Some of the stories I do not want to remain buried for ever in the museum files of dead magazine-numbers an author's not unpardonable vanity..."*    Fiction Pages 272

**Life of David Crockett** *by David Crockett*    ISBN: *1-59462-250-7*   **$27.45**
*"Colonel David Crockett was one of the most remarkable men of the times in which he lived. Born in humble life, but gifted with a strong will, an indomitable courage, and unremitting perseverance...*    Biographies/New Age Pages 424

**Lip-Reading** *by Edward Nitchie*    ISBN: *1-59462-206-X*   **$25.95**
*Edward B. Nitchie, founder of the New York School for the Hard of Hearing, now the Nitchie School of Lip-Reading, Inc, wrote "LIP-READING Principles and Practice". The development and perfecting of this meritorious work on lip-reading was an undertaking...*    How-to Pages 400

**A Handbook of Suggestive Therapeutics, Applied Hypnotism, Psychic Science**    ISBN: *1-59462-214-0*   **$24.95**
*by Henry Munro*    Health/New Age/Health/Self-help Pages 376

**A Doll's House: and Two Other Plays** *by Henrik Ibsen*    ISBN: *1-59462-112-8*   **$19.95**
*Henrik Ibsen created this classic when in revolutionary 1848 Rome.  Introducing some striking concepts in playwriting for the realist genre, this play has been studied the world over.*    Fiction/Classics/Plays 308

**The Light of Asia** *by sir Edwin Arnold*    ISBN: *1-59462-204-3*   **$13.95**
*In this poetic masterpiece, Edwin Arnold describes the life and teachings of Buddha.  The man who was to become known as Buddha to the world was born as Prince Gautama of India but he rejected the worldly riches and abandoned the reigns of power when...* Religion/History/Biographies Pages 170

**The Complete Works of Guy de Maupassant** *by Guy de Maupassant*    ISBN: *1-59462-157-8*   **$16.95**
*"For days and days, nights and nights, I had dreamed of that first kiss which was to consecrate our engagement, and I knew not on what spot I should put my lips..."*    Fiction/Classics Pages 240

**The Art of Cross-Examination** *by Francis L. Wellman*    ISBN: *1-59462-309-0*   **$26.95**
*Written by a renowned trial lawyer, Wellman imparts his experience and uses case studies to explain how to  use psychology to extract desired information through questioning.*    How-to/Science/Reference Pages 408

**Answered or Unanswered?** *by Louisa Vaughan*    ISBN: *1-59462-248-5*   **$10.95**
*Miracles of Faith in China*    Religion Pages 112

**The Edinburgh Lectures on Mental Science (1909)**  *by Thomas*    ISBN: *1-59462-008-3*   **$11.95**
*This book contains the substance of a course of lectures recently given by the writer in the Queen Street Hail, Edinburgh. Its purpose is to indicate the Natural Principles governing the relation between Mental Action and Material Conditions...*    New Age/Psychology Pages 148

**Ayesha** *by H. Rider Haggard*    ISBN: *1-59462-301-5*   **$24.95**
*Verily and indeed it is the unexpected that happens! Probably if there was one person upon the earth from whom the Editor of this, and of a certain previous history, did not expect to hear again...*    Classics Pages 380

**Ayala's Angel** *by Anthony Trollope*    ISBN: *1-59462-352-X*   **$29.95**
*The two girls were both pretty, but Lucy who was twenty-one who supposed to be simple and comparatively unattractive, whereas Ayala was credited, as her Bombwhat romantic name might show, with poetic charm and a taste for romance. Ayala when her father died was nineteen...*    Fiction Pages 484

**The American Commonwealth** *by James Bryce*    ISBN: *1-59462-286-8*   **$34.45**
*An interpretation of American democratic political theory.  It examines political mechanics and society from the perspective of Scotsman James Bryce*    Politics Pages 572

**Stories of the Pilgrims** *by Margaret P. Pumphrey*    ISBN: *1-59462-116-0*   **$17.95**
*This book explores pilgrims religious oppression in England as well as their escape to Holland and eventual crossing to America on the Mayflower, and their early days in New England...*    History Pages 268

QTY

**The Fasting Cure** *by Sinclair Upton*　　　　　　　　　　　ISBN: *1-59462-222-1*　**$13.95**
*In the Cosmopolitan Magazine for May, 1910, and in the Contemporary Review (London) for April, 1910, I published an article dealing with my experiences in fasting. I have written a great many magazine articles, but never one which attracted so much attention...* New Age/Self Help/Health Pages 164

**Hebrew Astrology** *by Sepharial*　　　　　　　　　　　　ISBN: *1-59462-308-2*　**$13.45**
*In these days of advanced thinking it is a matter of common observation that we have left many of the old landmarks behind and that we are now pressing forward to greater heights and to a wider horizon than that which represented the mind-content of our progenitors...* Astrology Pages 144

**Thought Vibration or The Law of Attraction in the Thought World**　　ISBN: *1-59462-127-6*　**$12.95**
*by William Walker Atkinson*　　　　　　　　　　　　　　　　Psychology/Religion Pages 144

**Optimism** *by Helen Keller*　　　　　　　　　　　　　　ISBN: *1-59462-108-X*　**$15.95**
*Helen Keller was blind, deaf, and mute since 19 months old, yet famously learned how to overcome these handicaps, communicate with the world, and spread her lectures promoting optimism. An inspiring read for everyone...* Biographies/Inspirational Pages 84

**Sara Crewe** *by Frances Burnett*　　　　　　　　　　　　ISBN: *1-59462-360-0*　**$9.45**
*In the first place, Miss Minchin lived in London. Her home was a large, dull, tall one, in a large, dull square, where all the houses were alike, and all the sparrows were alike, and where all the door-knockers made the same heavy sound...* Childrens/Classic Pages 88

**The Autobiography of Benjamin Franklin** *by Benjamin Franklin*　　ISBN: *1-59462-135-7*　**$24.95**
*The Autobiography of Benjamin Franklin has probably been more extensively read than any other American historical work, and no other book of its kind has had such ups and downs of fortune. Franklin lived for many years in England, where he was agent...* Biographies/History Pages 332

| | |
|---|---|
| **Name** | |
| **Email** | |
| **Telephone** | |
| **Address** | |
| | |
| **City, State ZIP** | |

☐ **Credit Card**　　　　☐ **Check / Money Order**

| | |
|---|---|
| **Credit Card Number** | |
| **Expiration Date** | |
| **Signature** | |

*Please Mail to:*　Book Jungle
　　　　　　　　　　PO Box 2226
　　　　　　　　　　Champaign, IL 61825
*or Fax to:*　　　　630-214-0564

## ORDERING INFORMATION

**web***: www.bookjungle.com*
**email***: sales@bookjungle.com*
**fax***: 630-214-0564*
**mail***: Book Jungle  PO Box 2226  Champaign, IL 61825*
**or PayPal** *to sales@bookjungle.com*

***Please contact us for bulk discounts***

## DIRECT-ORDER TERMS

**20% Discount if You Order
Two or More Books**
Free Domestic Shipping!
Accepted: Master Card, Visa,
Discover, American Express

www.ingramcontent.com/pod-product-compliance
Lightning Source LLC
Chambersburg PA
CBHW080057280326
41934CB00014B/3348